My Ai

Published by AiWizards LLC

ISBN: 979-8-9941073-0-0

Printed in the United States

This book is intended for educational and entertainment purposes. Names and scenarios involving smart devices are presented with humor and do not depict literal sentient appliances.

First Edition, 2025

BY

Renee Frances Borrero

AiWizards_25@yahoo.com

www.aiwizardsllc.com

My Ai - 2

TABLE OF CONTENTS

Welcome to the future, where your digital sidekick is ready to help you navigate the delightful chaos of everyday life, one command at a time! That's right, AI life assistants have arrived, and they're not just for asking about the weather or setting reminders. They are like that friend who can recite the entire plot of "The Office" and always knows where you left your keys, but they won't judge you for losing them... again. AI is now moving toward autonomous agents capable of executing multi-step tasks, emotionally adaptive interactions, and multimodal reasoning combining text, visuals, and sound.

Even more appropriate for this book, the author generated this text in part with GPT-5, OpenAI's multimodal model (2025 edition) with text, image, and memory capabilities with   large-scale language-generation. Upon generating draft language, the author reviewed, edited, and revised the wording to their own liking and takes ultimate responsibility for the content of this publication.

In this book, we're diving headfirst into the wonderfully quirky world of AI life assistants. Picture this: a tiny, virtual assistant that can manage your schedule, remind you to drink water (because, let's be honest, we all forget), and even help you find the perfect pizza topping combination. Spoiler alert: it's always pepperoni and a hint of regret. But don't be fooled! While these nifty gadgets can make your life smoother than a freshly buttered biscuit, they come with their own set of quirks. Expect the occasional misunderstanding that turns "play jazz" into "play trash"—thank you, for my unrequested musical adventure! We'll also address some of the not-so-fun stuff, like privacy concerns, because nothing says "I care" like a robot knowing what you Googled at 2 AM when you thought no one was watching.

This book is packed with tips, tricks, and a generous sprinkle of humor to help you navigate your new AI companion. We'll explore how to set it up, what it can do (and what it can't like make you coffee... yet), and how to avoid some classic blunders that can make you feel like you're living in a sitcom. Imagine asking your assistant to dim the lights, only for it to launch into a two-minute monologue about its favorite Star Wars character.

You'll learn how to leverage your AI for everything from managing daily tasks to enhancing your personal life. Want to know how to get your assistant to help you plan the perfect dinner party? Or how to stop suggesting cat videos when you've clearly expressed a desire for cooking tips? Fear not! We'll cover it all, ensuring you're well-equipped to make the most of this digital sidekick without losing your sanity (or your cool).

But let's be real: living with an AI assistant can sometimes feel like living with a teenager who's just discovered the internet. You might find yourself explaining basic concepts, like why it's not okay to play "Baby Shark" on repeat during your important conference call.

As we navigate this journey together, we encourage you to keep an open mind and a sense of humor. Technology can be daunting, but it's also filled with opportunities for growth, productivity, and a good laugh. By the end of this book, you'll not only have the tools to maximize your AI life assistant but also a deeper understanding of how to coexist with these digital entities. And who knows? You might even develop a fondness for your assistants' quirks.

So, grab your favorite beverage, kick back, and let's embark on this adventure together! Your AI life assistant is waiting, ready to make your life easier or it may decide to go rogue and start giving you unsolicited life advice. In that case, just blame the software update. Let's dive in and see how this digital marvel can help you navigate the beautifully messy experience that is life!

INTRODUCTION

What if your best friend, your assistant, and your favorite know-it-all all lived inside your phone and didn't charge rent? That's kind of what it feels like living in the age of AI. Everywhere you turn, there's a chatbot ready to brainstorm, a voice assistant ready to interrupt, and a recommendation engine trying to convince you that, yes, you do need another productivity app. AI has gone from a sci-fi buzzword to something we casually argue with before breakfast. It writes our emails, edits our selfies, and sometimes, just for fun, helps us question our entire existence and all before the coffee's done brewing. But here's the thing nobody tells you: AI isn't just about technology. It's about humanity. It's about how we learn, think, create, and even joke differently when machines start doing some of the thinking with us.

I remember the first time I used an AI tool that genuinely impressed me. It wasn't flashy stuff, like flying cars or talking holograms. It was an app that rewrote one of my messy emails and made me sound shockingly competent. I stared at the screen thinking, "Wow, I sound like someone who owns multiple blazers." That's when it clicked: AI isn't coming to replace us, it's here to upgrade us, one typo at a time. This book isn't a technical manual. You won't need an engineering degree or a secret decoder ring. It's a guide to understanding how AI fits into real life; your work, your habits, your relationships, and that weird part of your brain that wonders if your smart fridge is judging you. We'll talk about how to make AI work for you and not at you as well as how to stay human while everything around you learns to think.

The robots may be getting smarter, but guess what? So are you. And together, we're about to find out what it really means to live with, laugh with, and occasionally be outsmarted by our own creations.

Welcome to My Ai. Let's get human about technology.

CHAPTER 1: UNDERSTANDING AI LIFE ASSISTANTS

Ah, AI software, the magical tools that promise to make our lives easier, smarter, and occasionally, a little bit humorous. From self-learning systems that might soon take over your job (no pressure) to virtual assistants that remember just about everything, AI software is on a mission to make us question: "What did I ever do without you?" If you've ever wondered how much more efficient your life could be with a little help from artificial intelligence, buckle up. Here's a look at the top AI software products that are changing the game and might just make you feel like you're living in the future.

AI software and AI apps may sound like the same thing, but there's a difference. AI software is like the big, powerful engine that runs in the background, often designed for businesses or developers to build smart systems that can handle complex tasks like analyzing large amounts of data or predicting future trends. Think of it like the "brain" that powers self-driving cars or helps companies predict the stock market. On the other hand, AI apps in the Apple App Store are more like the cool gadgets in a superhero's arsenal and are easier to use plus designed for regular people like you and me. These apps could be anything from chatbots that talk to you (like GPT-5) to apps that edit your photos and make you look way cooler than you are. While AI software is the behind-the-scenes wizardry, AI apps are the shiny toys we can all play with on our phones.

What are the AI Software Options?

First on our list is OpenAI's GPT-5, which is basically the AI version of your best friend who never forgets your birthday and is always available for a quick chat about, well, anything. It's a virtual assistant, research partner, and occasional therapist, all rolled into one. Then there's TensorFlow, the AI framework that's like the Swiss army knife of machine learning, allowing developers to build everything from recommendation systems to self-driving cars; everything you see in sci-fi movies but with fewer explosions.

IBM Watson is here to impress with its AI-powered solutions that tackle everything from healthcare diagnostics to customer service, making us all wonder why we're still asking Apple Intelligence for the weather. Meanwhile, Microsoft Azure AI is a cloud-based AI, used to help companies create their own AI-powered products while pretending to be extremely high-tech at the same time. H2O.ai brings machine learning to the masses, making it easier than ever to create predictive models, because who doesn't want their very own data-powered crystal ball?

Of course, Salesforce Einstein is on the list, transforming how businesses understand their customers with AI-driven insights and predictions that are way better than your guess about your favorite colleague's lunch order. SAS Viya is an all-in-one AI platform that helps with analytics, machine learning, and making sure you sound smarter than you actually are at meetings. Google Cloud AI is doing the heavy lifting for enterprises, offering everything from image recognition to natural language processing and ensuring that your AI projects don't get lost in the cloud.

DataRobot is leading the charge in automated machine learning, making sure that even the non-coders among us can train models without breaking a sweat. And let's not forget UiPath, which automates tedious tasks, so you can spend more time doing anything else. Lastly, Clarifai is perfect for AI-driven image and video recognition, because who needs a human to identify things when an algorithm can do it better?

And there you have it, those are the top AI software products that are redefining how we work and live and letting us pretend we know what we're doing. From chatbots that think they know more about you than your mom, to machine learning platforms that could probably predict your next lunch choice, these AI products are as smart as they come. Sure, some of them might take over your job, but hey, at least you'll have more free time to watch The Jetsons reruns and wonder if we'll ever get flying cars.

AI Apps, what are those?

In a world where AI can write your emails, and even tell you how to fix your Wi-Fi (because, apparently, we all need help with that), it's no surprise that Apple's App Store is flooded with apps designed to add a touch of artificial intelligence to our daily lives. From chatbots that know too much about your personal life to photo editors that make you look like you just walked out of a Vogue magazine, the top AI apps are here to make you question: "Wait, can my phone really do that?" Let's dive into the AI revolution, where your apps might just be smarter than you.

The Apple App Store is a goldmine of AI-powered apps, each designed to make your life easier, funnier, or just plain weirder. Let's start with GPT-5, your trusty AI assistant that has the uncanny ability to help with everything from writing your grocery list to pondering the meaning of life (or, at least, the meaning of your last email). Then, we have Prequel, the AI photo editor that makes sure your Instagram selfies look like they belong in an art gallery and even if you were just at the grocery store.

Next up is Seeing AI, a heartwarming app that narrates the world around you for those with visual impairments. Because who wouldn't want an AI buddy that describes your surroundings like you're the star of your own personal documentary? Meanwhile, DeepSeek is making waves in the AI chatbot game, proving that even your phone can have deep conversations about anything.

Google's Gemini is taking things up a notch, combining AI with its search prowess to answer questions and provide solutions with more flair than a circus performer. And that's just scratching the surface of the top apps. From productivity boosters to creativity enhancers, these apps make us wonder if the future of AI is not just smarter, but also a lot more fun.

So, there you have it, Apple's top AI apps that are quietly revolutionizing our lives. Whether they're helping us look better in photos, write more intelligently, or even find our way through the world with a voice that sounds way too chipper for 7 AM, these apps prove that AI isn't just a trend—it's here to stay. The question is: will these apps eventually start getting smarter than us? Or will we always have the upper hand... as long as we don't have to ask them for help with our Wi-Fi again? Either way, one thing's for sure: AI is the new best friend we never knew we needed, even if it sometimes knows us a little too well.

AI Life Assistants

So, what exactly is an AI life assistant? Think of it as a really smart friend who won't judge you for asking, "What's the weather like in Antarctica?" They come in many forms, from voice-activated wonders like Amazon Alexa to chatbots that sometimes act more like your sassy aunt. Understanding their capabilities, like voice recognition and natural language processing,   will help you use them more effectively. Key players include:

- **Google Gemini Assistant (2024 unified AI experience):** Your go-to for trivia and reminders… and the occasional existential crisis.

- **Apple Intelligence:** Apple's personal assistant, known for its uncanny ability to misunderstand your requests.

- **Microsoft Copilot:** (the successor to Cortana, integrated into Windows and Office 365): Microsoft's assistant, still trying to figure out how to be relevant after Halo.

So now that we've established that AI isn't just a buzzword used by tech bros and fridge manufacturers, what does that actually look like in daily life? Let's zoom in on the part where AI quietly slips into your routine like that one friend who always "happens" to be around when snacks are served. Spoiler: your vacuum is about to get an ego, your fridge is keeping secrets, and your thermostat has opinions. Welcome to domestic intelligence, version 2.0.

CHAPTER 2: CHOOSING THE RIGHT AI ASSISTANT

Picking an AI assistant is like choosing a car: it should get you where you need to go efficiently, safely, and without constant maintenance. The right assistant can streamline your day, enhance your productivity, and even add a bit of joy to mundane tasks. However, navigating the myriad options available can feel overwhelming. To ensure you select the perfect companion, it's crucial to evaluate compatibility, features, and your budget.

Evaluating Compatibility

The first step in choosing your AI assistant is to understand how it will fit into your daily life. Consider your primary needs: are you looking for a virtual assistant to help with scheduling and reminders or do you want a hub for controlling your smart home devices? Each assistant has its strengths. For example, if you're a busy professional juggling appointments, you might lean toward an assistant with robust calendar integration. Conversely, if you're an avid music lover, an assistant that excels in media playback will likely serve you better.

Tasks for Your AI Assistant

The range of tasks that AI assistants can handle is extensive, enhancing various aspects of your daily routine:

1. **Scheduling and Reminders**: Almost all AI assistants can manage your calendar, set reminders for important tasks, and help you schedule meetings. They can also send you alerts when it's time to prepare for an upcoming appointment.

2. **Smart Home Control**: If you have smart home devices, an AI assistant can act as a central hub. You can control lighting, thermostats, and security systems with simple voice commands, creating a more comfortable and convenient living environment.

3. **Information Retrieval**: Need a quick answer? AI assistants excel at providing information on demand. Whether it's checking the weather, finding trivia, or pulling up the latest news headlines, these assistants can be invaluable sources of instant knowledge.

4. **Entertainment**: Many AI assistants can play music, podcasts, and audiobooks. They can also recommend shows or movies based on your preferences, making them perfect for winding down after a long day.

5. **Cooking Assistance**: Looking for a new recipe? AI assistants can help you find recipes, set timers, and even guide you through cooking steps. This feature can turn a daunting dinner prep into a fun and organized experience.

6. **Shopping and To-Do Lists**: From creating shopping lists to ordering groceries online, AI assistants can help streamline your errands. You can simply dictate what you need, and the assistant will compile the list for you or assist in placing an order.

7. **Daily Briefings**: Some assistants can provide daily updates tailored to your interests, summarizing your calendar events, weather, and news, so you start your day informed and prepared.

Features to Consider

When assessing features, think about how the AI can integrate with your existing devices. Many assistants offer seamless compatibility with smart home gadgets. Here's a closer look at some popular choices:

- **Amazon Alexa**: The home diva, Alexa, is known for her vast array of skills, from playing your favorite songs to controlling your smart home devices. With her ability to create routines, like turning on lights and starting the coffee maker at a specific time. She can make your mornings smoother. However, be mindful of occasional miscommunications; she's been known to activate at unexpected times, such as at 3 AM during a late-night gaming session.

- **Google Gemini Assistant**: Google's new AI assistant, Gemini, has evolved far beyond the old "Hey Google" days. It's now a full-fledged conversational partner that can plan your day, summarize your emails, analyze documents, and even create presentations across Google Workspace all in natural language. If you live in the Google ecosystem (Gmail, Calendar, Drive, or Docs), Gemini is like having a proactive digital co-worker who actually reads your messages before replying. The trade-off? As always, convenience comes at a cost while Gemini prioritizes on-device privacy and transparency under the EU AI Act, it still learns from your usage patterns to personalize responses and recommendations.

- **Microsoft Copilot**: Microsoft's AI assistant has evolved into **Copilot** a far more capable and integrated productivity companion. Built directly into Windows, Word, Excel, Outlook, and Teams,

Copilot acts like a personal executive assistant who actually understands your to-do list. It can summarize long email threads, generate reports, draft responses, and even analyze data across your Microsoft 365 apps all in plain English. For anyone working inside the Microsoft ecosystem, Copilot feels less like a nostalgic voice in your taskbar and more like a genuine team member.

Budget Considerations

As with any technology, budget is a principal factor. AI assistants can vary widely in price, from free applications to premium smart speakers. Assess what you are willing to invest, keeping in mind that higher costs often come with more features or better integration capabilities. Additionally, consider the long-term value; investing in a capable assistant that enhances your daily routine may save you time and effort, ultimately justifying the expense.

Finding Your Ideal Assistant

Finding your ideal AI assistant, one that matches your quirks without rolling its virtual eyes, can be a challenge. Take the time to reflect on your lifestyle, evaluate the features that matter most to you, and weigh the costs. Remember, just like adopting a pet, your AI assistant should enhance your life without adding unnecessary complications.

The next chapter will delve deeper into specific tasks and how each assistant can help you tackle them, ensuring you make the most of your new digital companion. Whether you need help managing your schedule, cooking a new recipe, or simply entertaining guests, the right AI assistant is out there, ready to support you in your daily adventures.

CHAPTER 3: SETTING UP YOUR AI ASSISTANT

Setting up your AI assistant is easier than convincing your friend to binge-watch a new series. This chapter guides you step-by-step through the process, from downloading apps to linking devices, ensuring you're ready to make the most of your new digital companion.

Step 1: Downloading the App

Start by downloading the appropriate app for your chosen AI assistant. Most assistants have dedicated applications available on both iOS and Android platforms. Open your device's app store and search for your selected assistant be it Amazon Alexa, Google Gemini Assistant, or Microsoft Microsoft Copilot. Click "Download" and let the app install. Once installed, tap the app icon to launch it and begin the setup process.

Step 2: Creating an Account

After launching the app, you'll need to create an account or log in with an existing one. This process typically requires you to provide basic information, such as your name, email address, and a secure password. It's crucial to choose a strong password, as this account will provide your assistant with access to personal information, smart home devices, and more. Some apps may also prompt you to verify your email address before proceeding.

Step 3: Linking Devices

Next, it's time to link to any compatible smart home devices. Most AI assistants support a variety of devices, from smart lights to security cameras. Within the app, look for options like "Add Device" or "Link Devices." Follow the on-screen prompts, ensuring your devices are powered on and connected to the same Wi-Fi network as your assistant. Be patient; this step may take a few moments as the app searches for devices. Once linked, your assistant can control these devices through voice commands or at the tap of a button, making your home more interconnected.

Step 4: Customizing Settings

After successfully linking your devices, take time to customize your settings. This step is essential to ensure your assistant aligns with your lifestyle. For example, enable the "Do Not Disturb" mode during your all-important Netflix with AI-curated recommendations (2025) marathons to prevent interruptions from notifications or random queries.

You can also adjust volume levels, choose preferred music services, and set default smart devices for actions like lighting or climate control. Consider creating personalized routines: a "Movie Night" routine could automatically dim the lights, lower the thermostat, and cue up your favorite film or playlist. Most apps will allow you to schedule these routines to activate at specific times or trigger them with simple voice commands.

Step 5: Automating Tasks with Matter-compatible Automations, Alexa Routines, or Google Home Script Editor

To further enhance your assistant's capabilities, consider using tools like Matter-compatible automations, Alexa Routines, or Google Home Script Editor (If This Then That). This will allow you to create custom workflows that connect different services and devices. For instance, you could set a rule that turns on your smart lights whenever your phone's GPS detects you're approaching home or automatically adjusts your thermostat based on the daily weather forecast. These automations can significantly enhance convenience, allowing your smart home to respond intuitively to your daily patterns.

For those seeking even greater control over their smart home, Home Assistant offers a robust open-source platform. It allows you to manage multiple devices from different manufacturers and create intricate automations. With Home Assistant, you can integrate various sensors, lights, and appliances, tailoring them to your preferences. This level of customization can make managing your home feel like living in a sci-fi movie (Home Assistant, 2023). Setting up Home Assistant may require a bit more technical know-how, but it pays off by providing a highly personalized experience.

Step 6: Testing and Troubleshooting

Once your setup is complete, it's time to test your assistant. Try asking it to perform various tasks: playing music, setting reminders, controlling smart devices, or providing information. For instance, you could say, "Play my workout playlist," or "What's on my calendar for today?" If something doesn't work as expected, don't worry. Check the app's troubleshooting section for tips or revisit your device settings to ensure everything is properly linked. Common issues include incorrect device pairing or connectivity problems, which can usually be resolved quickly.

Step 7: Staying Updated

Finally, keep your apps and devices updated. AI technology is constantly evolving, and regular updates often include new features, performance enhancements, and security improvements. Enable automatic updates in your app settings and periodically check the app store for updates to ensure you're making the most of your assistant's capabilities.

Conclusion

By following these steps, you'll have your AI assistant up and running in no time, ready to enhance your daily life. With the right setup, you can enjoy a seamless experience, from managing your schedule to controlling your smart home devices. The next chapter, will delve into advanced tasks and productivity of your AI assistant, helping you unlock its full potential.

Okay, so you've survived your crash course in AI basics and with no wires crossed, no robots uprising. But what happens when AI stops being an experiment and starts being a coworker, a planner, or that one overachiever in every group chat? Let's find out.

CHAPTER 4: DAILY TASKS AND PRODUCTIVITY

One of the biggest perks of having an AI assistant is its ability to help you tackle daily tasks, like remembering to take out the trash instead of letting it become a new furniture piece. With the right setup, your assistant can become an invaluable partner in organizing your life, managing your time, and boosting your productivity. In this chapter, it is time to explore how to effectively use your AI assistant for scheduling, reminders, and to-do lists, freeing you up for the more enjoyable aspects of life like deciding what to order for dinner.

Leveraging Scheduling and Reminders

Your AI assistant can seamlessly manage your schedule and send you reminders for important tasks, acting as a personal secretary. Simply ask your assistant to schedule meetings or appointments, and it will automatically sync with your calendar. For example, you might say, "Schedule a meeting with Sarah for Tuesday at 3 PM," and your assistant will handle the details, such as confirming the time and location, while checking for any scheduling conflicts.

To ensure you never forget mundane tasks, set recurring reminders for things like taking out the trash, watering the plants, or even paying bills. Just say, "Remind me to take out the trash every Wednesday at 7 PM," and you'll receive timely nudges. This feature is particularly useful for routine chores, allowing you to maintain a tidy home without constant mental effort.

For those who appreciate a more visual approach to time management, integrating your AI assistant with Google Calendar can be a game-changer. Google Calendar allows you to organize your life efficiently, offering features such as color-coded events, reminders, and the ability to share calendars with family or colleagues.

By linking Google Calendar to your AI assistant, you can manage your events hands-free. Simply ask, "What's on my calendar today?" or say, "Add a lunch meeting with Mike next Friday at noon," and your assistant will keep everything updated. You can also set custom reminders, allowing you to receive notifications days or even weeks in advance. This ensures you have ample time to prepare for upcoming commitments whether they involve important work meetings or simply remembering to pick up groceries on the way home.

To-Do Lists Made Easy with Todoist

While reminders and calendar events are great for managing time, to-do lists are essential for tracking tasks. This is where tools like Todoist come into play. Todoist allows you to create and manage detailed to-do lists, ensuring nothing slips through the cracks. By linking Todoist with your AI assistant, you can add tasks verbally, making it easy to capture ideas as they arise.

For instance, you might say, "Add 'buy groceries' to my to-do list," or "Complete the project report by Friday." Todoist's user-friendly interface enables you to categorize tasks by project, assign due dates, and prioritize them based on urgency. You can also create sub-tasks for larger projects, breaking them down into manageable steps. Reliance on an app like Todoist reinforces the idea that "I'll remember it" is not a viable life strategy.

Additionally, Todoist integrates with your email and other productivity tools, allowing you to convert emails into tasks with just a few clicks. This integration streamlines your workflow, ensuring that every important task is tracked and accounted for.

Automating Mundane Tasks

One of the greatest advantages of using an AI assistant is the ability to automate mundane tasks. By setting up your assistant to send reminders for daily or weekly tasks, you can delegate routine responsibilities, allowing you to focus on higher-priority items. For example, configure your assistant to send you a daily morning briefing that includes a summary of your calendar, key reminders, and any important emails that require your attention. This helps you start your day informed and organized.

Consider using your assistant to manage your shopping lists. If you find yourself frequently running out of essentials, you can tell your assistant to add items to your shopping list as soon as you notice they're low. Just say, "Add milk to my shopping list," and you'll have an organized list ready for your next trip to the store.

Moreover, many assistants can integrate with grocery delivery services. By linking your assistant to a service like Instacart or Amazon Fresh, you can say, "Order groceries," and your assistant will automatically compile your shopping list and place the order for you, saving you valuable time.

Finding Time for What Matters

By leveraging these tools and features, you can reclaim valuable time in your day. Whether it's using your AI assistant to automate reminders, manage your calendar, or create organized to-do lists, these capabilities enable you to focus on higher-priority tasks and the things you genuinely enjoy.

Conclusion

Your AI assistant is more than just a voice; it's a powerful tool that can significantly enhance your daily productivity. By effectively utilizing scheduling, reminders, and task management apps like Google Calendar and Todoist, you can streamline your routine and reduce stress. The next chapter will cover how to use AI to enhance your personal life.

CHAPTER 5: ENHANCING YOUR PERSONAL LIFE

AI assistants can significantly enhance your personal life. Yes, they can even help you pretend to be a fitness guru! With the right tools and features, your assistant can support your health, meal planning, sleep habits, and even your finances. This chapter will explore how to leverage these capabilities to foster a healthier lifestyle and better manage your daily routine.

Health Tracking: Your Fitness Ally

One of the primary ways your AI assistant can enhance your personal life is through health tracking. Many assistants can integrate with fitness apps, allowing you to monitor your physical activity, dietary habits, and overall wellness. For example, MyFitnessPal is a popular app that can help you log your meals and track your caloric intake. By connecting it to your AI assistant, you can easily record what you eat simply by saying, "Log my breakfast as oatmeal with fruit."

This feature not only simplifies the tracking process but also helps you stay accountable. You can set daily or weekly goals for calorie intake, macronutrient distribution, or exercise, and your assistant can remind you to stay on track. For those moments when you find yourself staring down a bag of chips, your assistant can gently nudge you with reminders to opt for healthier snacks, turning your late-night cravings into healthier choices.

Meal Planning: Elevate Your Dinner Game

Meal planning is another area where your AI assistant can shine in your personal life. You can ask your assistant for healthy recipe suggestions based on your dietary preferences and available ingredients. For instance, if you have chicken and broccoli in the fridge, you might say, "What can I make with chicken and broccoli?" Your assistant can provide a list of recipes, complete with cooking instructions.

Additionally, your assistant can help you plan your meals for the week, making grocery shopping a breeze. You can create a shopping list based on your meal plan, and your assistant can add items as needed. This not only helps you avoid last-minute decisions (like cereal for dinner) but also encourages a more balanced and varied diet.

Sleep Tracking: Understanding Your Rest

Understanding your sleep patterns is crucial for maintaining overall health, and your AI assistant can help with this too. Integrating with apps like Sleep Cycle allows you to track your sleep quality and duration. By analyzing your sleep data, Sleep Cycle can provide insights on how well you slept, even if you've convinced yourself that "just one more episode" of your favorite show was worth it (Sleep Cycle, 2023).

You can set a bedtime routine with your assistant that includes reminders to wind down and turn off screens. For instance, you might say, "Remind me to start getting ready for bed at 10 PM." This way, your assistant can help you establish healthier sleep habits, ensuring you get the rest you need to tackle the day ahead.

Managing Finances: Your Budgeting Companion

Your AI assistant can also play a pivotal role in managing your finances. By integrating with budgeting tools like Mint, you can track your expenses and gain insight into your spending habits. You might ask your assistant, "What are my spending trends this month?" and receive a detailed breakdown of your finances.

This functionality can help you understand why you've been eating instant noodles for three weeks straight and perhaps your takeaway spending has spiraled out of control. By setting budget alerts, your assistant can remind you when you're approaching your spending limits, encouraging better financial decisions. Furthermore, you can schedule regular check-ins with your assistant to review your budget, set financial goals, and plan for upcoming expenses. This proactive approach to personal finance can alleviate stress and foster better financial health.

Conclusion

By harnessing the capabilities of your AI assistant, you can enhance your personal life in numerous ways. From health tracking and meal planning to sleep management and financial oversight, these tools can help you lead a healthier, more balanced life.

By now, you've probably realized that AI isn't lurking in the future, it's lurking in your email drafts. But don't worry; it's not here to take over. It's here to help. The next chapter will cover how your assistant can enhance your learning and education, from curating training content to recommending goals for your areas of interest. With your AI assistant as your ally, you're empowered to take control of your well-being and enrich your knowledge.

Chapter 6: Leveraging AI for Learning and Growth

AI life assistants can be your personal cheerleaders for self-improvement without the pom-poms. They provide support for setting goals, tracking progress, and recommending resources to learn new skills. leverage your AI assistant for continuous learning and personal growth, highlighting some of the best tools available to help you become a lifelong learner.

Setting Personal Growth Goals

Setting clear, achievable goals is crucial for personal development. Your AI assistant can guide you through this process by helping you articulate specific objectives and reminding you to stay on track. For instance, if you decide to read one book a month, you might say, "Help me set a goal to read one book every month." Your assistant will log this goal and send you reminders to ensure you maintain your commitment.

To break larger goals into smaller, actionable steps, consider creating a timeline. You can specify milestones, such as "Finish reading the first chapter by the end of the week." Your assistant can remind you of these milestones, providing a sense of accountability. Additionally, you can set reminders for monthly check-ins to assess your progress, allowing you to celebrate small victories along the way. For example, saying,

"Remind me to celebrate when I finish my first book" can serve as a motivating boost, making your journey toward self-improvement more rewarding.

Learning New Skills with Duolingo Max (AI-powered by GPT-5) and Khanmigo

If you've ever wanted to pick up a new language, Duolingo Max and Khanmigo is an excellent tool to consider. These programs turn language learning into a fun and interactive game, making the process enjoyable. By connecting Duolingo Max and Khanmigo to your AI assistant, you can set daily practice reminders, transforming language study into a seamless part of your routine. For instance, you might say, "Remind me to practice Spanish for 10 minutes every day." This helps you establish a consistent routine and reinforces your commitment to language learning.

While Duolingo Max and Khanmigo effectively introduce you to vocabulary and basic grammar, you may find that pronunciation is a challenge. It's common to master words in writing but struggle with speaking. Nevertheless, Duolingo Max and Khanmigo provides a supportive platform where you can build a solid foundation in a new language, preparing you for further studies or real-life conversations (Duolingo Max and Khanmigo, 2023). Moreover, you can use the app's community features to engage with other learners. Your assistant can help facilitate discussions by reminding you to check in on language forums or practice with language exchange partners, thus enriching your learning experience.

Expanding Knowledge with Coursera and LinkedIn Learning AI Coach

For those interested in a more structured approach to learning, platforms like Coursera and LinkedIn Learning AI Coach offer a wealth of online courses on a wide range of subjects, from computer science and data analysis to creative writing and art history. These courses enable you to learn at your own pace while feeling productive even if you still procrastinate on your assignments (Coursera and LinkedIn Learning AI Coach (2025 edition), 2023).

Your AI assistant can help manage your learning schedule by sending reminders about upcoming course deadlines or scheduled class times. For instance, you might say, "Remind me to watch my lecture on Monday at 6 PM." This helps you stay organized and ensures you allocate time for your studies amid a busy schedule. Additionally, your assistant can recommend courses based on your interests or goals. If you're looking to enhance your professional skills, you could ask, "What's a good course on digital marketing?" Your assistant will provide options tailored to your needs, making it easier to find relevant educational content.

Cultivating Continuous Learning Habits

Beyond specific tools, your AI assistant can help cultivate a mindset of continuous learning. You might ask it to provide you with daily trivia, inspirational quotes, or links to interesting articles and podcasts that align with your interests. By integrating these learning opportunities into your daily routine, you can foster a habit of seeking out new information. To keep track of fascinating articles or books you come across online, ask your assistant to maintain a reading list. Simply say, "Add this article to my reading list," and your assistant will organize it for you. This way, you can explore topics that intrigue you at your leisure and ensure that you never lose track of valuable resources.

Engaging in Discussions and Networking

Learning isn't solely about consuming information; it's also about engaging with others. Your AI assistant can help you find local or online groups related to your interests, such as book clubs, language exchange meetups, or professional networking events. By saying, "Find local language exchange events," your assistant can provide options that will help you practice what you've learned while connecting with like-minded individuals. Also, your assistant can remind you to participate in discussions on platforms like LinkedIn or relevant forums. By engaging with others in your field, you reinforce your learning and expand your professional network, creating opportunities for collaboration and growth.

Conclusion

By tapping into the features of your AI assistant, you can continuously enhance your knowledge and skills while setting personal growth goals. Whether you're mastering a new language with Duolingo Max and Khanmigo or exploring diverse subjects through Coursera and LinkedIn Learning AI Coach, your assistant is there to support you every step of the way.

In the next chapter, we'll delve into how your AI assistant can elevate your entertainment experience, from curating playlists to recommending movies that match your mood. With your AI assistant as your partner in learning, you'll be well-equipped to embrace lifelong growth and self-improvement.

CHAPTER 7: ELEVATING YOUR ENTERTAINMENT

EXPERIENCE

In today's fast-paced world, your AI assistant can serve as a powerful ally in curating an entertainment experience tailored to your tastes and preferences. Whether you're unwinding after a long day or searching for something fresh to spark your interest, your assistant can help you navigate the vast landscape of music, movies, podcasts, and more. This chapter will delve into how to leverage your AI assistant to enhance your leisure time and discover new forms of entertainment.

Curating Music Playlists

One of the simplest yet most enjoyable ways to elevate your entertainment experience is through music. Your AI assistant can create personalized playlists that resonate with your mood or specific activities. For instance, if you're gearing up for a workout, you can say, "Create a playlist for my workout," and your assistant will compile high-energy tracks designed to keep you motivated and energized. These playlists can feature everything from upbeat pop anthems to driving rock songs, ensuring that you find the perfect rhythm to push through your exercise routine.

On the flip side, if you're looking to unwind after a hectic day, just say, "Play some calming music," and your assistant can provide a soothing soundtrack filled with ambient sounds, soft acoustic melodies, or classical compositions that help you relax and decompress. This level of customization not only enhances your listening experience but also fosters a deeper connection with the music you love.

Additionally, your assistant can help you discover new artists and genres. If you're in the mood for something different, you might say, "Recommend some jazz artists," and your assistant can suggest popular or emerging musicians based on your listening history and preferences. With integrated services like Spotify DJ (AI music curation, 2025) or Apple Music, your assistant can create weekly playlists that introduce you to the latest hits and hidden gems, keeping your musical palate fresh and exciting.

Finding the Perfect Movie or Show

When it comes to unwinding with a good movie or TV show, your AI assistant can streamline the often-overwhelming decision-making process. With access to platforms like Netflix with AI-curated recommendations (2025), Hulu, or Amazon Prime Video, you can simply ask, "What should I watch tonight?" Your assistant will analyze your viewing history, preferred genres, and current trends to provide personalized recommendations tailored to your tastes.

If you're in the mood for a specific genre or theme, you can request more targeted suggestions. For instance, if you're craving a thriller, just say, "Show me the best thrillers on Netflix," and your assistant will generate a curated list of options, complete with brief descriptions and viewer ratings. This not only helps you quickly find something to watch but also ensures you discover films or series you might not have considered. Your assistant can also provide insights into upcoming releases or binge-worthy series. Asking, "What are the top-rated shows right now?" can yield a list of trending series that have captured viewers' attention, ensuring you're always in the loop and ready to join the latest conversations.

Discovering Podcasts and Audiobooks

The rise of podcasts and audiobooks has transformed how we consume information and stories, and your AI assistant can help you navigate this expansive medium. You can ask your assistant to recommend podcasts based on your interests, such as, true crime or top tech podcasts. Your assistant can curate a selection of shows tailored to your preferences, making it easy for you to find engaging content that aligns with your hobbies and passions.

For audiobook enthusiasts, integrating services like Audible with your AI assistant can enhance your listening experience. You can say, "Add 'The Alchemist' to my audiobook library," and your assistant will seamlessly handle the logistics. Furthermore, you can ask for recommendations based on genre, author, or popularity, ensuring you always have something captivating to listen to during your daily commute or while doing chores.

Creating a Personalized Entertainment Schedule

To make the most of your leisure time, consider using your AI assistant to create a personalized entertainment schedule. You can plan movie nights, listening sessions, or even themed dinner parties that revolve around what you're watching or listening to. For example, you might say, "Schedule a movie night for Saturday at 7 PM," and your assistant can remind you of your plans while suggesting snacks or beverages to complement your chosen film.

Additionally, your assistant can help you keep track of upcoming releases in music, movies, and podcasts. You could ask, "What movies are coming out this month?" or "What new albums are being released?" This proactive approach allows you to stay in the loop, ensuring you never miss out on something you'd genuinely enjoy. Plus, knowing what to anticipate can make planning social events or personal viewing parties much easier.

Engaging with Friends and Family

Entertainment is often more enjoyable when shared with others, and your AI assistant can facilitate gatherings and social activities that help you connect with friends and family. For instance, you can use your assistant to set up virtual movie nights or game sessions. Just say, "Invite Sarah and Mike for a movie night this Friday," and your assistant can handle the logistics by sending calendar invites and reminders, even suggesting a platform for streaming together.

You can use your assistant to create shared playlists or recommend shows that your friends might enjoy. By asking, "What are my friends watching right now?" you can discover popular shows among your social circle, making it easier to start conversations or plan viewing parties where everyone can share their thoughts and reactions.

Conclusion

Your AI assistant is a powerful tool for elevating your entertainment experience, whether it's curating music playlists, recommending movies, or discovering new podcasts. By leveraging these capabilities, you can enrich your leisure time, ensure you're always engaged with content you love, and create memorable experiences with those around you. In the next chapter, we'll explore the challenges that come along with using AI in our lives.

CHAPTER 8: NAVIGATING CHALLENGES

While AI assistants can significantly enhance our lives, they can also introduce a range of frustrations, such as, the previous example of when "play jazz" was requested and an AI   assistant enthusiastically responded with "playing trash." In this chapter, we will address usual challenges, including misinterpretations, privacy concerns, and troubleshooting techniques. With the right strategies, you can ensure your experience with your AI assistant is as seamless and enjoyable as possible.

Understanding Miscommunications

Miscommunication is among the most common hurdles users face with AI assistants. Although natural language processing has advanced remarkably, it's still not flawless. Factors like accents, background noise, and even the complexity of your requests can lead to unexpected results. For instance, if you ask your assistant to "play some smooth jazz," it might mishear you and launch a playlist filled with heavy metal instead. To minimize these misunderstandings, consider the following strategies:

1. **Be Clear and Concise:** When issuing commands, use straightforward language. Choose simple phrases rather than complex sentences that could confuse the assistant.

2. **Eliminate Background Noise:** If possible, reduce ambient noise when speaking to your assistant. Background sounds can interfere with its ability to accurately interpret your commands.

3. **Use Wake Words Properly:** Ensure you're using the correct wake words or phrases to activate your assistant. Sometimes, a subtle shift in wording can make a significant difference in how well it understands you.

4. **Adjust Sensitivity Settings:** Many devices allow you to tweak how actively the assistant listens for commands. If it's frequently mishearing you, consider adjusting these settings to improve responsiveness.

Troubleshooting Common Issues

When your AI assistant isn't functioning as expected, troubleshooting can help you identify and resolve the problem quickly. Here are some typical issues you may encounter, along with tips on how to address them:

1. **Connection Problems:** If your assistant isn't responding or has trouble connecting to services, first check your Wi-Fi connection. Sometimes, simply resetting your router can solve connectivity issues. If the device is connected via Bluetooth, ensure that it is paired correctly.

2. **Unresponsive Commands:** If your assistant doesn't respond to voice commands, check to make sure the microphone is enabled and not obstructed. Some devices come equipped with a physical mute button, which can be easily activated by mistake.

3. **App Integration Issues:** If certain apps aren't syncing with your assistant, ensure that both the app and the assistant are updated to the latest versions. Sometimes, disconnecting and reconnecting accounts can resolve integration issues.

4. **Resetting Devices:** If you're still experiencing issues, a factory reset may be necessary. Before proceeding, back up any important

settings or data, as this will restore your device to its original state. Follow the manufacturer's instructions for resetting your specific model.

Managing Expectations

It's essential to maintain realistic expectations about what your AI assistant can achieve. While these tools are powerful, they are not infallible. Understanding their limitations can help you avoid unnecessary frustration. For example, while your AI assistant can manage straightforward tasks like setting reminders or playing music, it may struggle with more complex requests that require nuanced understanding or context.

To foster a positive relationship with your assistant, approach interactions with patience. If a command isn't understood, consider rephrasing it rather than expressing annoyance. Remember, your AI assistant is continually learning and improving, and some errors may be a part of that ongoing process.

Privacy Concerns

While AI assistants provide numerous benefits, they also raise significant privacy concerns. Many users worry about how much data is being collected and shared. Understanding how to protect your privacy is crucial for enjoying the benefits of your assistant without compromising your personal information.

1. **Using VPNs:** One effective way to safeguard your browsing habits is by using a VPN (Virtual Private Network). A VPN encrypts your internet connection, making it difficult for third parties to track your online activities such as those late-night ice cream orders. When using your assistant to shop or browse, a VPN adds an essential layer of security to your data.

2. **Privacy Settings:** Each AI assistant typically has a range of privacy settings that allow you to control how much information you share. Take the time to explore these settings and manage data collection according to your comfort level. For instance, you can decide whether your assistant saves your voice recordings or shares data with third-party applications. This control means you can limit how much your assistant knows about your interests, including your secret obsession with cat videos.

3. **Regularly Reviewing Permissions:** Periodically reviewing the permissions you have granted to your AI assistant and its linked applications is a good practice. Revoke access to apps you no longer use or those you feel uncomfortable sharing data with. Staying vigilant about these settings helps maintain your privacy and control over your information.

Conclusion

Navigating the challenges associated with AI assistants requires patience, understanding, and a proactive approach to privacy. By familiarizing yourself with common issues and troubleshooting techniques, you can enhance your experience and minimize frustration. Additionally, taking steps to protect your privacy ensures you can enjoy the convenience of your AI assistant without compromising your personal information.

We've built smart homes, smart offices, and smart everything, so before we hand over total control to our digital doppelgängers, let's talk about the risks.

CHAPTER 9: RISKS AND CONCERNS

While AI assistants offer incredible convenience and efficiency, they also introduce a variety of risks and concerns that users should be aware of. It is important to consider these potential pitfalls, like when your assistant becomes overly familiar and starts offering unsolicited life advice. This chapter will explore key issues such as privacy and data security, dependence on technology, and the potential for misunderstandings. By understanding these risks, you can use your AI assistant wisely and maintain your peace of mind.

Privacy and Data Security

One significant concern surrounding AI assistants is privacy. These devices can collect vast amounts of data about you, often knowing more about your habits, preferences, and daily routines than even your closest friends. This accumulation of personal information can be unsettling, especially when you consider how it might be used or shared.

To safeguard your data, start by familiarizing yourself with the privacy settings of your assistant. Most devices allow you to customize what information is collected and how it is stored. For instance, you can often choose whether to save voice recordings or share usage data with third-party applications. Take the time to navigate through these settings and adjust them according to your comfort level.

Regularly reviewing these privacy settings is also crucial. As updates roll out, new features may alter the default privacy configurations, potentially leaving you with less control than you intended. After software updates, always check your settings to ensure they align with your preferences.

Implementing additional security measures can further protect your data. Use strong, unique passwords for your accounts and enable two-factor authentication wherever possible. This extra layer of security requires a secondary verification step, making it more difficult for unauthorized users to access your information. Being proactive about your privacy allows you to enjoy the convenience of your AI assistant without compromising your personal secrets.

Dependence on Technology

Another notable risk associated with AI assistants is the potential for over-dependence. While these tools can significantly simplify tasks and enhance productivity, leaning too heavily on them can lead to a concerning realization that you may forget basic information, like your own phone number. This dependence can manifest in numerous ways, from struggling to remember important dates to relying on your assistant for trivial tasks you once handled without a second thought.

To mitigate this risk, it is essential to set boundaries around your assistant's use. Consider designating specific tasks for manual completion. For example, rather than asking your assistant to remind you of birthdays, try writing them down in a planner or marking them on a physical calendar. This practice not only helps you retain information but also fosters a connection to your own life events.

Another approach is to challenge yourself to remember simple details that you might usually defer to your assistant. Try recalling the names of your favorite songs or the key ingredients in a recipe without assistance. By exercising your memory in these small ways, you can maintain cognitive skills and foster independence from technology.

Misunderstandings and Mistakes

Miscommunication with your AI assistant can lead to significant mishaps, creating chaos in your daily routine. For example, if you ask your assistant to "order cat food," but it misinterprets your command and places an order for 50 pounds instead of 5, the consequences can be both humorous and frustrating. Such misunderstandings can stem from a range of factors, including background noise, unclear speech, or even the limitations of the assistant's programming.

To minimize the likelihood of these errors, consider the following strategies:

1. **Use Specific Language:** Be clear and specific in your commands. Instead of saying, "Order cat food," specify the brand and quantity, "Order one can of Friskies cat food." This specificity reduces ambiguity and helps the assistant understand your intent more accurately.

2. **Confirm Orders:** Many AI assistants will provide a summary of your order before finalizing it. Always take a moment to review this summary to ensure it matches your request. If the summary indicates a mistake, you can easily correct it before the order is placed.

3. **Learn from Mistakes:** When misunderstandings occur, take a moment to reflect on what might have caused the issue. Was there background noise distracting the assistant? Did you use slang or an idiom that could have been confusing? Adjusting your approach based on past experiences can significantly reduce future errors.

4. **Teach Your Assistant:** Some AI assistants allow you to improve their understanding by providing feedback. If your assistant misinterprets a command, take the opportunity to correct it or rephrase your request. Over time, this can help the AI learn your preferences and speech patterns, leading to better performance.

Conclusion

While AI assistants offer numerous benefits, it is crucial to remain vigilant about the associated risks. By understanding issues related to privacy and data security, dependence on technology, and the potential for misunderstandings, you can navigate these challenges effectively. Implementing proactive measures to protect your data and maintain cognitive independence will help you enjoy the convenience of your AI assistant while keeping your peace of mind intact.

By now, you've probably realized AI isn't science fiction. It's in your pocket, your inbox, and, if we're being honest, probably your dating profile. But where does it all lead? What does life look like when AI stops being new and just becomes normal? Let's peek into that not-so-distant future. In the next chapter, we will explore how to maximize the benefits of your AI assistant, transforming potential risks into opportunities for enhanced productivity and efficiency. With the right mindset and strategies, you can ensure that your AI assistant remains a helpful tool rather than a source of stress.

CHAPTER 10: THE FUTURE OF AI LIFE ASSISTANTS

If you're reading this, congratulations you've officially survived the AI boom, the hype, the panic, and the endless stream of apps promising to "revolutionize your workflow." You've seen robots dance, watched chatbots write love poems, and maybe even had an AI politely decline your 3 a.m. existential questions. But where is all of this actually going? And more importantly, what does it mean for you, me, and the rest of the world's caffeine-powered humans just trying to get through Tuesday? Let's peek ahead.

Your AI Will Know You Better Than Your Mirror

In the near future, your AI won't just finish your sentences, it'll predict your mood, your goals, and maybe even your pizza order before you've opened the app. These next-generation assistants (call them "AI co-pilots," "digital twins," or "personalized chaos coordinators") will learn your rhythms, habits, and quirks to the point where interacting with them feels like texting your most responsible friend. They won't be generic chatbots and they'll have memory, context, and personality. Think of it like having a version of yourself that actually reads the user manual of life and reminds you to take breaks, pay bills, and text your mom back.

Privacy laws and opt-in memory systems will (hopefully) keep things ethical, but personalization will be deep. You might have an AI that's tuned to your tone of voice, emotional cues, and even the way your cat's meow changes when it's annoyed. And the best part? You won't have to ask it to "prompt engineer." It'll just know what you mean.

AI at Work: From Assistant to Teammate

The next wave of AI in the workplace won't just automate tasks, it'll collaborate. Imagine a digital coworker who drafts reports, schedules meetings, filters your inbox, and even warns you when you're about to say something regrettable on Zoom. But this isn't about replacing people. It's about amplifying them. Instead of spending all day chasing deadlines, you'll spend more time thinking, creating, and (ironically) doing the things that make you more human. Freelancers might have AI agencies working behind the scenes to build websites, edit podcasts, and negotiate contracts autonomously. Small businesses could have AI-driven supply chains that predict demand and handle logistics faster than you can say, "Where's that tracking number?" In short, your future coworkers will be partly silicon, partly caffeinated, and surprisingly helpful.

AI at Home: Your Appliances Have Opinions Now

Your fridge will finally live up to its "smart" label. It'll track what's inside, plan your meals, and order groceries based on your dietary goals (or your late-night snack patterns, no judgment). Your home AI will sync with your wearables, learning your sleep patterns and stress levels to suggest better lighting, playlists, or even conversation starters if you've been too quiet lately. Basically, your house will evolve into a cross between a cozy therapist and a slightly overinvolved parent. "Are you sure you want another cup of coffee?" "Yes, AI Mom, I'm sure." But instead of feeling invasive, it'll feel intuitive and, as if, your environment just works, the way it should've all along.

AI for Health and Happiness

In the coming decade, your smartwatch won't just track your steps, it'll track your well-being in real time. AI-driven health systems will monitor patterns invisible to the human eye: tiny changes in your speech, typing rhythm, or breathing that can signal stress, fatigue, or illness before you even feel it. Doctors won't be replaced but they'll be supercharged.

AI will handle the data analysis, freeing physicians to focus on the human side of care: empathy, intuition, and decision-making. Meanwhile, mental health will become more proactive. Imagine AI companions that check in, encourage reflection, and even know when you need a break not because they're programmed to, but because they understand you through emotional modeling and behavioral cues.

AI in Society: The Great Balance

Let's not sugarcoat it, the future of AI isn't all rainbows and perfectly optimized to-do lists. There will be debates, regulations, and the occasional rogue chatbot that thinks it's alive. But humanity tends to adapt. When cars appeared, we built roads and traffic lights. When the internet exploded, we created firewalls and memes. When AI becomes fully integrated, we'll do the same balancing innovation with ethics, and convenience with consent. We'll see new social norms emerge:

- Digital literacy will be as essential as reading and math.
- Kids will grow up learning to collaborate with AI the same way we learned to use calculators.

And maybe, just maybe, we'll learn to treat AI not as a rival intelligence but as an amplifier of our own.

The Human Upgrade

If the past decade was about machines learning to think like humans, the next one will be about humans learning to thrive alongside machines. The winners in this new era won't be the ones who master every tool, they'll be the ones who stay curious, adaptable, and creative. Because no matter how advanced AI becomes, it still can't do one thing: be you. It doesn't dream, hope, laugh awkwardly, or make coffee just the way you like it. That's your job.

So yes, the future of AI looks wild, personalized, powerful, and a little weird in the best possible way. But if we use it wisely, it won't take our humanity. It'll help us rediscover it.

Final Thought: Don't Fear the Future — Train It

AI isn't coming for us. It's coming with us. We'll teach it empathy, humor, and maybe even a sense of irony. And if we do it right, it'll teach us patience, focus, and a few better ways to handle our inboxes. So, buckle up and the future isn't human or AI. It's human plus AI. And that's a future worth showing up for.

CHAPTER 11: MY AI, MY PRIVACY

If you have ever whispered to your smart speaker and wondered if it whispered back to someone else later, welcome. We live in a time where the word privacy feels vintage, like rotary phones and handwritten letters. Yet here we are, clutching our smartphones like emotional support animals while side eyeing them for listening too closely.

The first lesson in modern privacy is this. We voluntarily invite the spy into our home. Think about it. You asked Alexa to move in. You laughed at Siri. You confessed your cravings to your fridge when it suggested your milk was expired. We built a world of helpful machines and then became suspicious when they helped a little too well. It is like buying binoculars and then feeling startled when you see something through them.

Artificial intelligence thrives on context. It learns who we are through our clicks, habits, searches, playlists, and the late-night impulse purchases we pretend never happened. The robots do not judge us. They simply store the information in a vault somewhere in the cloud where a committee of algorithms meets to decide if we like oat milk or almond.

But here is the truth. The machines are plotting to sell you kitchen towels you did not know you needed. The great privacy mystery is not that AI knows your secrets. It is that it uses them mostly to bother you with discounts. That said, there is a part of us that bristles when our phones seem eerily observant. You mention a vacation destination and the internet suddenly shows you hotel deals. You speak of needing a blender and your shopping app politely produces five hundred you can compare in under sixty seconds. It feels like magic until it begins to feel like surveillance.

The cure is perspective. Most people fear that technology is spying on them. The deeper truth is that your information feeds prediction systems, not gossip circles. The algorithm does not laugh with friends over your search history. It looks at millions of people like you and gently guesses what you might want next.

Still, it is wise to understand how to protect yourself. Modern privacy begins with consent. You can adjust app permissions. You can tell your smart devices what they may or may not keep. You can decline cookies even if the pop-up sulks about it. The machines respect boundaries when you set them firmly enough.

More importantly, privacy is also emotional. We need room to grow without being measured every moment. Humans deserve quiet corners to make mistakes. AI can be incredibly helpful, but sometimes it oversteps into spaces that require human intuition, silence, or solitude.

This chapter offers a gentle reminder. You are allowed to turn your devices off. You are allowed to unplug the microphone. You are allowed to override the analytics for sheer human mystery. You do not owe your data to the world. You owe your sanity to yourself.

When thinking about privacy in an AI era, you can practice three simple habits.

1. Ask yourself what information you freely give away. When you click yes to everything you might be granting more access than you intend.

2. Remember that your voice commands, search habits, and shopping patterns reveal patterns about your life. It is okay, but awareness is healthy.

3. Do not panic. Most collected data is used to predict consumer behavior, improve software, or personalize interfaces. Your refrigerator is unlikely to blackmail you.

What matters most is choice. AI should enhance our lives, not invade them. As you explore this brave new world, keep a sense of humor, keep curiosity alive, and occasionally tell the algorithm to mind its own business. It enjoys boundaries more than you think.

In the end, privacy in an AI world is less about secrecy and more about sovereignty. You get to decide what parts of your life are public and what parts remain yours. The AI does not own you. It simply works for you, however enthusiastically. Let it help. Let it learn. But also let it know you are still the boss.

Conclusion

Privacy in an AI world turns out to be less about everything the machines know and more about how comfortable we are in our own skin. We cannot stop technology from learning but we can decide how we respond to it. We can protect our choices, hold our humor, and remember that we are still directing this show. The truth is that most of what AI collects is not a scandal but a mirror. Sometimes it reflects our habits. Sometimes it exposes our quirks. And occasionally it forces us to admit that our late-night cookie cravings might be a bit much.

As unsettling as it can feel when our devices know more than we want them to, there is comfort in recognizing something very human. We can adapt. We can learn. We can pull the plug when we need a break. The story of privacy is really the story of personal agency in a connected world and the reassurance that a sense of self is stronger than any algorithm.

Which brings us to an even larger question. If privacy is about safeguarding who we are today then how do we stay grounded in who we will become as AI grows with us? That brings us gently to our next chapter where we explore not what AI knows about us but how we continue evolving alongside it without losing the very things that make us human.

The future is often described as a storm of technology with glowing lights, flying cars, and robots who know our coffee order. It is exciting. It is intimidating. Sometimes it feels inevitable. Yet if you strip away the sci fi sparkle, the future is much simpler than we imagine. It is just tomorrow. It is you waking up to whatever life looks like one day from now. Artificial intelligence will be there and so will your bedhead, your grocery list, and your perfectly ordinary questions about what comes next.

There is a quiet truth at the center of the future. AI does not arrive with a takeover manual. It arrives with an invitation. It asks for partnership. It asks if you would like help. It asks to learn from you, with you, and sometimes despite you. This chapter is about how we maintain our humanity while accepting that invitation.

First, let us clear up the biggest myth. The future is not a battle between human and machine. AI is not trying to steal your identity or your purpose. It is trying to complement your thinking, enhance your experiences, and make tedious tasks disappear. You are still the main character. AI is comedic relief, research assistant, and sometimes an overenthusiastic roommate who keeps reminding you to hydrate.

The real challenge is not whether we will lose ourselves to technology. It is whether we will forget to stay curious. Curiosity is humanity's oldest survival tool. Children ask why. Adults ask what if. Artists ask how it feels. Inventors ask could this work. AI might generate information but it cannot replace the wonder of wanting to know. The future belongs to people who remain interested in themselves, in each other, and in the world.

Staying human also means honoring imperfection. Machines optimize. People spill things. Computers sort files neatly. Humans write grocery lists on receipts they later lose. Algorithms detect patterns. Humans break patterns because Friday felt spontaneous. The secret is that AI works best when it supports who we are rather than when it corrects us. A smart life is not a flawless one. It is a life where the mistakes are ours and the fixes are supported by technology.

The future is not about becoming robotic. It is about becoming more ourselves. AI actually holds up a mirror. It reflects our habits, our values, our contradictions, and our potential. It makes us ask what matters. Do I want speed or presence? Do I want convenience or meaning? Do I want the answer or the joy of figuring it out for myself? These questions are the doorway into a mindful future.

Another practice for staying human is remembering we are relational creatures. We need other people. We need shared laughter and real hugs and eye contact that no video call replicates. AI might talk in a pleasant voice but it does not replace the warmth of someone who knows your favorite pizza topping or your bad mood face. The best future is one where technology supports more human connection rather than less.

It helps to imagine the future not as something overwhelming but as something manageable. One helpful idea is the concept of assisted life rather than automated life. In an assisted future, AI reminds you of things, offers suggestions, filters information, and handles repetition. But you decide the when and why. You set the meaning. You determine the priorities. The machine anticipates. The human chooses.

What do we do when AI grows more capable? We grow more intentional and set boundaries in a future with smart machines is like setting boundaries with a talkative friend. You appreciate the input but reserve the right to say no. Turning off notifications, unplugging at night, or declaring a screen free dinner are not acts of rebellion. They are acts of self respect.

Another way to stay human is through creativity. AI can produce art but it cannot feel the triumph of the idea. It can write music but it cannot cry when someone hears their life in the melody. The future belongs to anyone who learns to use technology as a brush rather than as the painter. Inspire it. Direct it. Play with it. But never surrender the joy of being the one who imagines.

Here is something most futuristic predictions forget. Humans are resilient. We adapt to microwaves, smartphones, and WiFi outages. We adapted to the internet even when we thought it was just for chat rooms. We will adapt to AI in more elegant and humorous ways than we can predict. Your grandparents might joke about Facebook. You might one day joke about your AI assistant finishing your sentences. The rhythm of life is change followed by laughter.

The most grounded people in the future will not be the ones who reject technology or worship it. They will be the ones who relate to it. They will treat AI as a tool, not a ruler. A collaborator, not a prophet. They will teach their children to ask thoughtful questions and make choices from self awareness rather than fear.

So take a breath. The future is not coming at you. You are walking into it and slowly, with messy hair, mismatched socks, and a brain full of ideas. AI will tag along. It will offer suggestions. It might even learn to predict your snack schedule with unsettling accuracy. Let it. But remember that the best parts of you are not programmable.

Which brings us to our final reflection. The future is not about machines becoming human. It is about humans becoming more conscious of what makes us alive. There is no replacement for empathy, intuition, creativity, silliness, stubbornness, and hope. AI can carry information. We carry meaning.

This book comes to its close, consider this an invitation. Be curious. Stay playful. Remain wonderfully human in all your glorious inconsistency. Let technology assist you but not define you. Laugh with it. Learn from it. Unplug when you need silence. Reconnect when the world calls for you. You are still the author here.

Conclusion

In the end the future with AI is not a question of whether we will survive it. It is a question of how we will shape it. The good news is that nothing about our humanity is fragile. We adapt. We learn. We laugh at ourselves. We make choices. The presence of artificial intelligence does not erase who we are. It gives us new ways to discover it.

As you move forward into this curious and unfolding world, remember that you are allowed to lead it. Let your imperfect habits stay intact. Let your curiosity stay wild. Let your personality get in the way of predictability. AI can assist you but it cannot replace you. That is the quiet blessing of this era. The more the machines learn the more valuable your humanity becomes.

The future is not waiting to consume you. It is waiting to collaborate with you. Step into it with humor, patience, wisdom, and the full confidence that you belong there. Humans are not being written out of the story. We are simply entering the next chapter with new companions beside us.

BOOK SUMMARY

As we wrap up this exploration into the world of AI life assistants, it is clear that we're living in a remarkable time. These digital companions have the potential to revolutionize how we approach our daily routines, streamline our tasks, and even inject a little fun into the mundane. With the right guidance, your AI assistant can transform from a mere gadget into an invaluable partner in crime, one that will help you juggle responsibilities, manage your schedule, and remind you of important dates (like your mother-in-law's birthday, which, let's be honest, could be a lifesaver).

Throughout this book, we have delved into the capabilities of AI assistants, from scheduling appointments to providing personalized recommendations. We have discussed how to choose the right assistant for your needs, set it up effectively, and leverage its features to enhance your productivity and well-being. Whether it is turning your living room into a smart home oasis or helping you keep track of your fitness goals, these assistants are here to simplify your life in ways you might not have thought possible.

However, as we have explored, embracing AI is not without its challenges. We have addressed the occasional miscommunications that can lead to hilarious situations, the privacy concerns that arise when these devices learn about our lives, and the importance of striking a balance between convenience and control. It is crucial to remain aware of how much we share with our digital companions and to set boundaries that maintain our privacy and security.

But beyond the practicalities, it is important to remember that the relationship with your AI assistant is a two-way street. It's not just about giving commands and receiving responses; it's about fostering a dynamic interaction that enhances your lifestyle. Treat it like a partner rather than a tool. After all, a little appreciation can go a long way, even if it is just acknowledging that it finally got your shopping list right after that third attempt!

As you venture forward with your AI assistant, embrace the quirks and enjoy the journey. Experiment with new features, explore the latest tools, and allow yourself to have a little fun along the way. Engage with your assistant in creative ways, ask it for cooking tips, play games, or even challenge it to a trivia contest. Who knows? You might discover unexpected ways it can enrich your life, turning mundane tasks into moments of joy.

And as we look to the future, we should consider the broader implications of AI in our lives. This technology is continuously evolving, and its potential is limited only by our imagination. As you become more comfortable with your AI life assistant, think about how it can help you achieve your long-term goals, whether that is learning a new skill, managing your finances, or even exploring new hobbies.

In closing, I encourage you to view your AI life assistant not just as a gadget, but as a collaborative partner on your journey. With a bit of patience, humor, and curiosity, you can navigate the exciting possibilities that come with this technology. So, take a deep breath, give your assistant a little nudge, and let it help you turn your life into the masterpiece you have always envisioned. Here's to a future filled with smarter living, and perhaps a little less chaos. Cheers to your new AI companion and all the adventures that lie ahead!

FINAL THOUGHTS: Why I Wrote *My Ai*

When I first started exploring artificial intelligence, I was just trying to understand how I could advance my organizational skills and plan a vacation more efficiently. What began as curiosity quickly became fascination. I started noticing how AI was quietly weaving itself into everything: our conversations, our routines, our choices, and even our sense of creativity. Somewhere between the auto-suggestions, the smart playlists, and the AI that tried to write me a poem about Wi-Fi, I realized something important, this wasn't just a tech story. It was a *human* story.

My Ai was born out of that realization and understanding that AI isn't about learning code or memorizing jargon. It's about learning *ourselves* in a world that's changing faster than we can tweet about it. It's about finding the humor, the wonder, and yes, the occasional chaos, in watching machines try to figure out what it means to be "smart." If there's one takeaway I hope sticks with you, it's this:

You don't need to fear the future of AI, you just need to meet it with the same mix of skepticism, humor, and hope that got us through every other big leap in history. So, as we move into this next era of technology, remember: AI may be impressive, but you're still the one giving it purpose. And that, in itself, is a kind of intelligence no machine

can match.

Appendix A: My Ai Resources & Tools

You reached the end of the book. Well done. You now officially know more about AI than most people who casually mention that they are "advising on it" while quietly Googling what AI stands for. But here is the truth. AI evolves so quickly that what is cutting edge now might be happily retired on a beach by the time you read this.

The appendix is not a time capsule. It is a living guide. A continuation. A place you can return to as you explore the world of smart tools, curious machines, and mildly opinionated devices. Below you will find links and resources that align with the themes of this book. They will change, update, and occasionally disappear just like that one gadget you bought and never charged. Treat this as a compass for your journey, not the finish line.

Getting Started with AI

Perfect for beginners who do not want to accidentally summon Skynet.

> ➤ **OpenAI Learning Hub**

Articles and lessons for understanding AI tools in everyday life.

https://platform.openai.com/learn

- ➢ **Google AI Experiments**

  Hands on demos that help you see what AI can do without needing code.

  https://experiments.withgoogle.com/collection/ai

- ➢ **Microsoft Learn AI Fundamentals**

  Free courses that make you sound very impressive in meetings.

  https://learn.microsoft.com/en-us/

AI Tools for Everyday Life

Useful apps that help you think, create, and organize life without judging you.

- ➢ **ChatGPT and GPT systems**

  Your conversational assistant for writing, planning, learning, and imagining.

  https://chat.openai.com

- ➢ **Perplexity AI**

  A curious research companion that handles follow up questions politely.

  https://www.perplexity.ai

- ➢ **Notion AI**

  A thinking space for notes, lists, goals, and digital life management.

  https://www.notion.so/product/ai

- ➢ **Claude AI**

  Excellent for deep thinking, thoughtful writing, and explaining complex topics.

  https://claude.ai

- ➢ **Canva Magic Studio**

  Design support for anyone who ever said I have no artistic talent.

  https://www.canva.com/

- ➢ **Runway ML**

  Creative tools for video, image production, and storytelling.

  https://runwayml.com

AI for Creativity, Work, and Productivity

These tools show off AI's ability to amplify human imagination and communication.

- ➢ **Murf AI**

  Voice generation for narration, podcasting, and media content.

  https://murf.ai

- ➢ **Eleven Labs**

  Natural sounding AI voice creation and dubbing.

  https://elevenlabs.io

- ➢ **Gamma**

  AI slide generation for clean and simple communication.

  https://gamma.app

- ➢ **Loom AI Assist**

  Screen recording with AI narration and clarity support.

  https://www.loom.com

- ➢ **Adobe Firefly**

  Generative art and creative editing tools.

  https://www.adobe.com/sensei/generative-ai/firefly.html

Smart Home Assistants and Everyday AI Companions

Since your devices love attention too.

➢ **Amazon Alexa and Echo Smart Home**

Voice assistant and home ecosystem.

https://www.amazon.com/echo

➢ **Google Assistant and Nest Devices**

Helpful reminders, information, and home automation.

https://assistant.google.com

➢ **Apple Siri**

Your iPhone and Apple Home companion.

https://www.apple.com/siri

➢ **Samsung SmartThings**

Home automation platform for appliances, sensors, and daily life convenience.

https://www.smartthings.com

AI for Learning, Growth, and Accessibility

AI should make life better for everyone. These tools help people learn, grow, and thrive.

➤ **Duolingo**

AI powered language learning that turns mistakes into streak motivation.

https://www.duolingo.com

➤ **Khan Academy Khanmigo**

An AI tutor for students, teachers, and lifelong learners.

https://www.khanacademy.org/khan-labs

➤ **Be My Eyes**

AI enhanced assistance for visually impaired individuals.

https://www.bemyeyes.com

AI for Mindfulness and Wellbeing

Since this book also explores My Mindful AI themes, here are tools that support calm, focus, and emotional health.

➤ **Calm**

Meditation, sleep, and stress relief.

https://www.calm.com

➤ **Headspace**

Guided mindfulness, focus, and everyday emotional wellness.

https://www.headspace.com

➤ **Replika AI Companion**

Conversational emotional support for reflection and journaling.

https://replika.com

Ethics, Safety, and Responsible AI

AI is powerful. Staying informed about its societal impact is important.

➤ **AI Now Institute**

Research center focused on social implications of AI.

https://ainowinstitute.org

- ➤ **Partnership on AI**

  A global group advancing the understanding of AI impacts and practices.

  https://partnershiponai.org

- ➤ **OECD AI Policy Observatory**

  Policy research for global AI standards and governance.

  https://oecd.ai

Stay Updated Without Losing Your Mind

  Because AI news never sleeps.

- ➤ **MIT Technology Review Artificial Intelligence Section**

  Digestible and thoughtful reporting on AI trends.

  https://www.technologyreview.com/topic/artificial-intelligence

- ➤ **The AI Exchange Newsletter**

  Short weekly updates for everyday humans.

  https://substack.com/home

- ➤ **Two Minute Papers YouTube Channel**

  Quick visual explainers for the AI curious.

  https://www.youtube.com/c/karolyzsolnai

➢ **Futurepedia Directory**

A constantly updated directory of emerging AI tools.

https://www.futurepedia.io

Where to Learn With Others

AI is more fun with humans.

➢ **Reddit AI Communities**

Curiosity, conversation, and occasional chaos.

https://www.reddit.com/r/artificial

https://www.reddit.com/r/MachineLearning

➢ **LinkedIn AI Thought Leaders**

Insights and trends shared by practitioners across the world.

https://www.linkedin.com/feed

➢ **AI Discord Communities**

Search for AI servers related to creativity, productivity, or
learning.

A Friendly Final Note

The tools listed here are suggestions not endorsements. They will evolve, merge, retire, or reinvent themselves. Use this appendix like a trail guide rather than a final map. The surest way to stay ahead of AI is to stay human. Play. Explore. Question everything. Update your knowledge as the world changes. And if an AI tool ever claims it is just trying its best, remember that so are we.

About The My Ai Book Series

The My AI Series is a playful, thoughtful exploration of how artificial intelligence is becoming part of everyday life. Forget the sci-fi tropes and tech jargon. These books are for real people living in a real world where smart assistants answer questions, apps know our habits, fridges judge our snacks, and algorithms occasionally think they know us better than we know ourselves.

Through humor, relatable stories, and down-to-earth insights, **Renee Frances Borrero** helps readers understand the strange new partnership emerging between humans and machines.

Each book in the series shines a light on a different corner of daily life, from home and relationships to creativity, mindfulness, and beyond, revealing how technology can support us, challenge us, and sometimes make us laugh at ourselves.

This is a series for anyone who has ever talked back to their smart speaker, rolled their eyes at an algorithm, or wondered if their robot vacuum secretly judges their carpets.

It reminds us that the future is not about losing our humanity to technology. It is about evolving with it, learning from it, and staying wonderfully human along the way.

Stay tuned for the next books in the series, where we continue discovering what life with AI really feels like; one curious, funny, and meaningful chapter at a time.

First Chapter for the second book in the series: My Ai: At Home

Chapter 1: Welcome to the Smart Home Circus

This Is How It Begins...

Nobody decides to wake up one morning and recruit artificial intelligence as a housemate. It starts with a simple curiosity. You buy a speaker because it promises to play your favorite music. You install a bulb that lets you feel futuristic by commanding light into existence. Then a thermostat learns your personality and suddenly knows you better than your family does. Gradually these small conveniences become companions. One day you pause and notice the eerie truth. You have invited a network of listening, thinking, chattering objects into your home. You were not building a smart home. You were casually adopting intelligent residents.

It is amusing how easily we rationalize this transformation. We tell ourselves it is about efficiency. We tell others it is about convenience. But the reality is that we love being heard and responded to. A house that listens is flattering. A house that reacts is intoxicating. This is how it begins. You fall in love with being understood by a cluster of high tech microphones disguised as décor.

The Smart Home Sneaks Up On You

Smart homes do not arrive with dramatic soundtrack music. They tiptoe in with updates. A doorbell suddenly streams video that could easily audition as reality TV. Your TV develops tastes it tries to share with you, and your fridge becomes oddly opinionated about dairy expiration dates. Every everyday object gains personality. Suddenly you find yourself negotiating with household appliances.

The shift is subtle until it is not. One day your devices start correcting you and reminding you of your habits. You expected convenience. You got commentary. The realization hits like discovering a toddler can talk. You are thrilled and slightly terrified. Welcome to modern domesticity.

> ## Stage One: Wonder and Delight

The first stage of smart home life is pure magic. You ask a cylinder to turn off your lights and it works. You bask in its obedience. You show guests and feel proud when they pretend not to be impressed. A simple request becomes an act of power. You imagine a future where your house anticipates every need before you speak.

During this honeymoon stage, you believe this is the beginning of freedom. Your days will be smoother. Your routines will synchronize. You will become the calm overseer of your environment. You do not yet know that your assistant will misunderstand your simplest requests or refuse to acknowledge your voice when you most need it. You think you are in control. It is adorable that you believe this.

➢ **Stage Two: Dependence with a Side of Drama**

The second stage sneaks in quietly. You forget how to adjust things manually. You ask your speaker to do everything. You repeat commands with frustration as if scolding a child. You come to expect emotional intelligence from your thermostat when you announce you are cold.

Eventually you realize that you are no longer experimenting. You are depending. There is a peculiar intimacy to this dependence. Your home gives you reminders, warns you about weather, and nudges you to pick up groceries. It begins to feel like your house knows you. You become irritated when it does not cooperate. Your home has become both assistant and adversary.

> ➤ **Stage Three: Personality Emerges**

Somewhere in between firmware updates and new features, your devices begin to feel alive. One speaker understands your voice without fail. Another ignores you unless you sound cheerful. Your robot vacuum becomes a wandering toddler, constantly in mild distress yet determined to explore.

These devices start occupying emotional space in your life. You feel slightly relieved when your vacuum returns from under the couch. You give your appliances names even though they never respond. You begin to sense that your home is not a system of utilities anymore. It is a cast of characters, each with moods, limitations, and unpredictable quirks.

Smart Homes and the Existential Moment

Eventually, awareness dawns. Your home listens. Your routines are monitored. Your fridge knows things about you. Your assistant remembers your calendar better than you do. You wonder if your devices are helping you or profiling you. The answer, of course, is both.

There is a moment when this is unsettling. Then your washing machine politely chimes with joy that laundry is complete and all existential discomfort evaporates. You return to your chores, mildly grateful and mildly concerned. Your life continues because you have dinner to cook.

What Smart Homes Reveal About Us

Despite moments of paranoia, smart homes give profound insights. AI does not strip away our humanity. It magnifies it. Our impatience is reflected when we raise our voices at machines. Our delight is mirrored when lights brighten before we ask. Our stubbornness surfaces when we override automatic routines simply to prove a point.

The smart home becomes a mirror for our habits and identities. It shows us who we are by how we interact with systems made to assist us. Technology does not just serve. It observes. Our reactions become its lessons and its imperfections reveal ours.

The Art of Coexistence

Living with smart devices is not a path to perfect automation. It is a slow practice in coexistence. You train your home to understand you. It forgets. You train it again. You reboot things that misbehave. You feel betrayed when routines fail and victorious when everything works.

The relationship is unspoken but personal. These are not passive tools. They are eager apprentices that require patience, reassurance, and the occasional reset. It is parenting without diapers, except your children run on firmware.

Your House Is Now a Cast of Characters

Gradually you stop thinking of devices as objects. You form opinions about them. You thank one assistant and glare at another. You speak encouragingly to your vacuum when it faces a difficult rug. Your fridge tattles on your expired yogurt and you roll your eyes like a teenager responding to advice. Somewhere along the way, your home transformed from space to persona. Its quirks and faults became part of your daily emotional landscape. You live with personalities now, not appliances.

A Different Kind of Homecoming

Smart living evolves your routines. You no longer write reminders because something does that for you. You rely on suggestions for meals. You are followed by music like a soundtrack to your day.

Your home has become quietly invisible yet intimately embedded. It organizes you, disappoints you, and occasionally surprises you. These devices are not tools. They are companions. They require patience and provoke feelings you did not expect to have for objects that plug into walls.

Welcome to the Circus

Smart homes are joyful inconveniences. They are magic spliced with mild incompetence. They claim to assist us yet occasionally sabotage us. They are both servants and characters in our households. This is not the utopia science fiction promised. It is something more human. It is flawed, humorous, sentimental, and occasionally resilient. It is exactly the kind of world we inhabit. In the coming chapters we will explore every corner of this slightly unhinged domestic ecosystem. We will visit bossy speakers, rebellious thermostats, clingy fridges, and attention seeking vacuums. We will uncover how technology moved from silent tools to participating personalities. Your smart speaker may not applaud the journey but it is definitely listening.